DEDICATION

For Milt, who shared in my passion for this place
and then made me sit still and write about it: thank you.

ADVANCE PRAISE FOR CHEWING SAND

"This memoir is the story of falling in love with a boundless desert landscape and the fierce effort to protect it from destruction. ...*Chewing Sand* will awaken the reader to that which seems unlovable. It is haunting, and causes one to see in new ways."

—Dana Greene, author and Dean emeriti of Oxford College of Emory University

"Gail Collins-Ranadive's contemplative adventures in the Nevada desert, amid brilliant rocks and human foibles, would provide a year of sermons or an activist's agenda. This book will be a real morale booster to those who focus on global warming. A latter day Transcendentalist, Gail's poetic presence gives us 'nature's scripture.' Here's to those who step aside for marigolds!"

—Peter Tufts Richardson, author of *Sunday Meditations and Four Spiritualities*

"In her collection of essays, the Reverend Gail Collins-Ranadive reveals the connection between the timeless and temporal, the spiritual and the profane, as she observes them in the Mojave Desert around Las Vegas. In the process, she makes herself one of the leading voices in the modern eco-spiritual movement."

—Will Moredock, author of *The Banana Republic: A Year in the Heart of Myrtle Beach*

"...*Chewing Sand* is a love song not only for the desert, but also for the earth as a whole, and draws the reader into the deep connection between spirituality and justice."

—Lynn Ungar, author of *Bread and Other Miracles*

"A beautiful anthem of gratitude. Gail has a reverence for her adopted desert's preciousness that echoes the sentiments of John Muir who was, likewise, a crusader for the conservation of natural resources...."

—Jackie Maugh Robinson, poet; Sun City, Summerlin Writers' Group

CHEWING SAND

An Eco-Spiritual Taste of the Mojave Desert

For Rose,
with gratitude
and admiration,
Blessings,
Gail
Collins Ranadive
June 2014

CHEWING SAND

An Eco-Spiritual Taste of the Mojave Desert

Gail Collins-Ranadive

PUBLISHED BY HOMEBOUND PUBLICATIONS

Homebound Publications books may be purchased for educational, business, or sales promotional use. For information please write: Special Markets Department, Homebound Publications, PO Box 1442, Pawcatuck, CT 06379

Visit us www.homeboundpublications.com

FIRST EDITION

ISBN: 978-1-938846-24-3 (PBK)
Book Designed by Leslie M. Browning
Cover Images by © Gert Hochmuth | Shutterstock.com

Library of Congress Cataloging-in-Publication Data

Collins-Ranadive, Gail, 1944-
Chewing Sand : an Eco-Spiritual Taste of the Mojave Desert / Gail Collins-Ranadive.—First edition.
pages cm
ISBN 978-1-938846-24-3 (pbk.)
1. Mojave Desert. 2. Self-consciousness (Awareness) 3. Ecocriticism. I. Title.
PS3553.O476328C48 2014
814'.54--dc23

2013049883

10 9 8 7 6 5 4 3 2 1

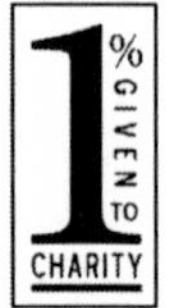

Homebound Publications holds a fervor for environmental conservation. We are ever-mindful of our "carbon footprint". Our books are printed on paper with chain of custody certification from the Forest Stewardship Council, Sustainable Forestry Initiative, and the Programme for the Endorsement of Forest Certification. This ensures that, in every step of the process, from the tree to the reader's hands, that the paper our books are printed on has come from sustainably managed forests. Furthermore, each year Homebound Publications donates 1% of our annual income to an ecological or humanitarian charity. To learn more about this year's charity visit www.homeboundpublications.com.

Contents

"In the desert there is everything and there is nothing. Stay curious. Know where you are—your biological address. Get to know your neighbors—plants, creatures, who lives there, who died there, who is blessed, cursed, what is absent or in danger or in need of your help. Pay attention to the weather, to what breaks your heart, to what lifts your heart. Write it down."

—Ellen Meloy

Something Missing

Looking down the sloping rock-strewn trail towards where a creek is supposed to be, I see…I see…I see… nothing.

Utterly, and absolutely, nothing…unless you call the vast, bleak, barren, sun-baked wasteland of shrub studded sand, that endless sand spreading everywhere, something, which I do not. I am a product of the verdant East, where I much prefer to be.

In fact, I'd deliberately returned home to New England after seminary in Berkeley to do my parish internship outside of Boston, and then taken a part time settled ministry on the south shore of Massachusetts with the expectation that it would grow to three-quarter and finally into a full time position. But financial crisis had followed financial crisis, and within two years the congregation had started spending down its meager endowment in order to pay my compensation package. So I reduced my hours in an effort to make things work for as long as possible. But wanting to do more ministry, not less, I knew I'd have to leave sooner rather than later, despite the efforts of one congregant to pay me in chocolate in the hope that I'd get too fat to get out of the church door.

Still wishing to stay in the archetypal landscape of my Transcendentalist Unitarian forebears, I'd put my name into the Interim Ministry pot and requested that it be sent primarily to church openings in New England. Yet in spite of phone conversations exchanged with various interim search committees, the process had been unbearably slow. So I'd sought solace out in the woods newly blooming with lady slippers that were picked to near extinction in my childhood.

Because my spiritual life has always been grounded in the natural world, I promised the powers that be: "I'll go anywhere—as long as there are trees!"

So I was unprepared that day I came in from my woods to find a message on my answering machine. A woman's voice announced that she was calling from Las Vegas, Nevada, and asked if I'd consider doing an interim year out there.

Clearly god had not lost her sense of humor: I hate heat, don't gamble, and need trees! I laughed for the three hours until, "Ms. Nevada" called back. Then I blurted out, "Do you have trees out there!?"

Now here I am, squinting in the bright light, straining to see if the shimmer of heat at the bottom of the hill truly is the line of pine trees for which this desert trail has been named.

It is only because the Las Vegas congregation maintains this trail that my search committee host has brought me out here, where a brutal July sun burns my back....my own fault for wearing relentless clergy black, even in this short sleeve shirt.

The intensity of the heat fries my mind as I stumble back toward the air-conditioned car. I take one final glance at the nothingness over my shoulder: clearly I am missing something!

Promised Land

Before I left MA, a friend gave me a post card entitled "The Promised Land." It showed a couple in a red convertible stopped beneath two signs: one sign pointed to Heaven, the other to Las Vegas. Standing beside the car is the white-robed figure of Jesus giving directions; he is pointing to Las Vegas.

This begs the question: promise of what? The postcard conflates three responses: the heart's desire, ones ego need with its shadow of greed, and the soul of the whole Self.

Because my spiritual path led me into the metaphoric desert of the Old and New Testaments, I now felt called to follow the Joshua Trees out into this desert. These spiny yuccas are the hallmark of the Mojave, and are the twisted apparitions that the early Mormon settlers named for the prophet who finally led the Israelites into the Promised Land.

Perhaps I would seek out the desert tradition of Moses and Jesus, the Magi and Mohammed, the early Christian desert Mothers and Fathers. But first things first.

Standing in a line that stretched the length of the Department of Motor Vehicles before it even opens, I join the 6000 people who've moved to this Promised Land this month.

As I surrender my old identity, I inquire about how to pronounce the name of my new state: I get to keep the A sounds in Massachusetts and transfer them both into Nevada!

To celebrate, I turn right out of the parking lot, heading away from the infamous Strip and up toward the Conservation Area, to purchase my annual pass as a brand new Nevadan.

At first, the surroundings suggest Anywhere America with its familiar shopping centers, gas stations and supermarkets. Residential areas are newer than those back in New England, yet many have similar lush lawns and towering trees, as if pretending this isn't a land of little rain and too much sun.

But suddenly *civilization* ends, and scrubby shrubs stretch as far as the eye can see. Then ancient creosotes rise up to claim swaths of space by clearing circles of sand so that nothing else can grow and thus threaten their water supply. Soon yucca and cholla give way to the beckoning Joshua Trees, which I follow further westward, out to the thirteen-mile loop of scenic road through preserved public land.

Thoreau wrote in *Walking* that we travel east to study history, literature, and art, "retracing the steps of the race," but we go "westward, into the future." And Emerson once claimed that we Americans needed the boundless West in order to become ourselves, to stop being pseudo-Europeans.

But for that we needed our own text, a new one that reflected this desert landscape of both spirit and experience.

I'd taken my paperback of *Desert Solitaire* to read at the DMV. Now Ed Abbey's words came alive as I saw: "…life not crowded upon life, but scattered abroad in spareness and simplicity, with a generous gift of space…"

This was truly an introvert's Promised Land! Out here there was no neon and noise, no bright lights so it was never night, no manically designed casinos promising continual stimulation, free drinks, beeping slots, and no visible way out.

For an introvert, is the promise one of time and space for an inner life, provided by the desert that would be my Refuge?

Silence

I have always craved silence. I credit this to growing up as the eldest of eight in a household where there was precious little peace and quiet. So during my nursing school rotation in a state psychiatric institution, where a patient earned spending money by painting wooden signs for our desks, my classmates commissioned plaques that carried their future married names. For mine I chose a singular word: Quiet. I have it still.

Thus I looked forward to the profound silence that the desert so famously promises, and drove out to the national conservation area just west of Las Vegas whenever I could.

Passing the outcrop of solidified sand dunes, where people of today were climbing up on the calcified remains of some of the earliest creatures on the earth, and appearing right sized as specks on the limestone of evolution, I would head for the highest point in the park.

Up there the spaciousness of place challenged my make do Massachusetts' mindset. With the wide horizon blessedly bounded in all directions by a ring of mountains, I could almost get my bearings. But beyond a single juniper bush, the desert opened as if to Infinity, and still bore the scar of the Old Spanish and Mormon Trails that were traveled by adventurers.

Yet I was to stay put for a year. And up here the silence was so deep that it seemed to swallow my heartbeat. Would I get lost in all this expansiveness? My knees felt like gelatin.

Blessedly, the place was also a popular tourist attraction, with cars and tour buses and SUV's and motorcycles bringing the valley's noise and air pollution with them: idling engines, blaring radios, and blowing cigarette smoke frequently disrupted the natural quiet. My ego self breathed in relief.

I think I know why we humans are so afraid of silence that we must keep it out there, at bay, away from us: its emptiness reflects our own inner void, that space we attempt to fill with self-destructive behaviors or deny with addictive consumerism.

But what is it that we don't want to know that we already intuitively know? And would that knowing move us out of the familiar security of what Thoreau called our "lives of quiet desperation," the desperation that comes, for instance, from being cooped up on an overbooked airplane, stalled in congested traffic, or caught in any crowded place that provokes confusion's outbursts of rage and violence?

Perhaps were we to let it, the desert's silence would grab us by the throat, and take our breath away, leave us standing still in awe and gratitude, beholding and wondering and feeling at one with, rather than apart from, the rest of creation.

Acquiring a Taste

Shaped by the landscape celebrated by my hometown poet Robert Frost, and aching for the colors of fall, I found my way to the cut off for Willow Spring Creek, where a cluster of yellowed cottonwoods held onto the last of the autumnal light.

For although I create and conduct worship services that nourish others, I feed my own spirit out in the natural world.

Yet out here there were just a few of trees, beyond which lay a dirt trail leading into the dust of the desert, with a yellow surprise of still blooming flowers scattered across the sand.

A posted sign read, "Take only pictures, leave only footprints." So being careful to stay on the trail, I walked until I came upon a series of red handprints splayed across the face of a beige rock, paintings made by the prehistoric peoples who frequented this area perhaps as early as 10,000 years ago.

Five hauntingly small hands seemed to be hugging the boulder, touching and touching in with the numinous power of the planet as it manifested itself in this particular landscape.

The reverence of these ancient ancestors wrested my attention away from the figurative desert of my childhood religion, and challenged me to pay attention to this actual one.

And more: I found myself wondering whether indeed, as Vine Deloria has written, on this continent, "God is Red."

This insight intrigued me. Did bringing their middle eastern religion with them blind both the pilgrims and pioneers, prevent them from bonding with this land, thus turning it into the abstract idea found in our musical lyrics, such as in "my country tis of thee, sweet land of liberty," or "this land is your land," or "all over this land," or "the land of the free."

Our patriotic allegiance to this country seems to have no relationship to the continent upon which live and have our being. In contrast, the Native peoples still experience this land as the source of all life, which is why they, "Return thanks TO our mother, the earth, which sustains us, TO the rivers and streams, which supply us with water, TO all herbs, which furnish medicines for the cure of our disease, TO the corn, and to her sisters, the beans and squashes, which give us life."

Was this the difference between an ego-self and an eco-self? Was being here an opportunity to consciously commune with this American desert, connect with its essential Nature, and learn from the wisdom that is native to this landscape?

Yet I knew that I must not misappropriate the life-ways of another culture. I must use the techniques of my own tradition.

This landscape would become my sacred text for contemplation (taking a long loving look at what's real), then for reading everything I could find about deserts in general and this desert in particular, then processing it all via writing, and finally for extending my experience through conversation with companions I'd hike with on this path and many other trails.

The five red handprints on the beige boulder beckoned me back here again and again, to where I could feel the anxiety and confusion of my extraverted profession drop away, and invited me to pay attention to what wasn't there until I looked again,...until I began acquiring a taste for this Place.

Coyote Duality

Out in the arroyo where I walked when I didn't have time to drive up to the Canyon, one day I heard the domestic dogs barking furiously in the yards lining the open space area.

Looking for a cause, I spotted a sand colored animal frantically searching for a way out of Pueblo Park.

But walls and fences lined the ridge of the wash, where, when it was not raining, people could enjoy walking along a paved path amid the natural fauna of the desert, with prickly pear and creosote bushes sheltering rabbits and lizards.

Now this coyote, an archetype of the desert southwest, was trapped in an environment made by and for humans.

I watched in fascination while this tawny dog with the disposition of a cat antagonized its domesticated cousins.

How had this sleek and slender creature become such a Rorschach test for people? Whether perceived as an intruder or a trickster says more about the human than the animal.

As I stood there transfixed, I too kept projecting my own sense of duality, for I was trying to choose between two paths of ministry. Interim or settled careers offered nearly opposite opportunities, and now my ambivalence was playing out right in front of my eyes. Might this even be the same coyote I'd seen drinking freely from a spring out in the Canyon? Out there it wore its wildness like a birthright. Here it seemed confounded.

The wild and the domestic, Father Thomas Berry assured us, are the two main forces of the universe: that expansive energy of the big bang counteracted by gravity's containment.

And every being contains both in its wholeness. The gift of a human lifespan seems to be the chance to embody more of one than the other at different times in our own narrative: sometimes our task is to live within the containing frame; at other times, we're invited to expand beyond our comfort levels and be challenged to trust in the wildness within.

I would eventually opt for the insecurity of interim ministry, always uncertain there would be a next position, then annually packing and moving for eight years, until the grief of perpetual leaving caught up with me, and called me to stay put in one place. Not surprisingly, I chose this place, where I first saw those two coyotes that may have been one and the same.

Trickster! For even as I was choosing the freedom of expansiveness, I was subtly gravitating towards this place through a longing as plaintive as the calls of a covey of quail on parade across the suburban street and into the open desert.

This strange attraction never let up, and never let me go.

Now I suspect that it began that same winter when I was interviewed for an article in the Las Vegas Sun about churches and their role in environmental protection.

Mine was the only congregation in town that was even minimally ecologically aware, so I was featured along with Joshua Abbey who was working within his Jewish tradition.

When he told us that his dad would have written off Las Vegas long ago as, "the Californication of Southern Nevada," his statement set off something deep within me.

And a fierce sense of dog-like loyalty to this severely compromised ecosystem began to take hold on me and grow.

I simply didn't know it at that time.

Water, Water...

A refrain from my childhood, the one about having enough sense to come in out of the rain, is upside down here, where we come in out of the relentless sunshine and, at the first sight of clouds, run outside to look up at the sky.

But, because of high temperatures and low humidity, much of our rainfall evaporates before reaching the ground.

Thus the rarity of rain makes us pay attention to the marvel of water, and we mindfully take it with us everywhere in refillable bottles. I learned this important lesson about desert living first hand when I got caught in the infamous spaghetti bowl interchange during evening rush hour in 104-degree heat.

Perhaps this is what it takes: feeling your mind become numb and your skin dry out helps you truly know that you are indeed made up mostly of water. And water becomes sacred.

No wonder the major world's religions have always taught that in the beginning was not the word but water, from the Hebrew story of the spirit of God moving over the face of the waters, to the Hindu story of the World Egg floating upon the primal ocean, from which the entire cosmos was born.

Our own personal beginnings parallel the cosmic one. We are made up mostly of water, we formed in water, and we gushed forth from water. Plus water is used to wash away sins, to purify, to anoint with new life, to baptize, and to symbolize rebirth...from Hinduism and Buddhism to Islam and Christianity.

For as Loren Eiseley notes, what are we humans being but a clever way that water has found for going beyond the reach of ponds. And so with the water within me longing for itself with any sign of rain, on cloudy days I open my door to sniff for the creosote that releases a pungent odor when it gets wet.

Because the valley's so vast, it can be raining in one place and not in any other, creating the perfect recipe for rainbows. Thus near the end of my interim year, as the late summer monsoons moved in, I drove to a high point and gazed in awe at a pair of double rainbows embracing the entire Vegas valley.

Rainbows have the mythic/mystical role of carrying messages between earth and the heavens, such as Yahweh's to Noah after the Flood. But I was seeing rainbows *before* the deluge hit my car! Then buckets of water dumped down on me.

I hadn't seen the wall of water coming until its torrents engulfed me, cutting off my view of anything beyond the car's dashboard. I fumbled for the windshield wipers, turning every knob I could find, one at a time: nothing worked.

The wipers just stayed still. Were they broken?!

No. After nearly a year without rain, I had simply forgotten where they were!

Out of the car, I was drenched, as a blessing, or was it a promise? No wonder the Swahili word for rain is the same as the word for *grace*! For like grace, we can't grasp it with our fist; rather, we have to let it flow into and through our cupped hands; we can only be gracefully, and gratefully, open to it.

Reality Check

We have no business being here! That reality hit me like a blast of desert heat as I headed down the interstate towards Las Vegas. I always knew I'd come back: I didn't know how, or when, only why: I felt called to go deeper into what I'd so briefly glimpsed during my interim ministry year in the desert.

Completing my drive across the continent, I was on the long stretch of *real desert* beyond the artificially created golf course greenery of Mesquite, NV. Creosote bushes dotted the landscape, each claiming an area around itself by poisoning the soil so that competitors couldn't steal the little water available.

My heart sank with an unwelcome thought: we the people do not belong here in this extreme environment! What WAS I doing returning to this area that was so totally unsustainable? How had Vegas become such a sprawling city of over 2 million people living in the hottest, driest desert in North America?

Was it all a cruel joke, one that denied reality? Had we intelligent people been sold a bill of goods as ludicrous as the earlier settlers who moved West with the assurances that, "rain followed the plow." Was I here as a witness to the final days left or, at most, to a decade of viability in this harsh landscape?

The American mindset set in motion by the idealists and industrialists back in the Nineteenth Century had become embodied in this valley, which prided and promoted itself as a place where anything is possible. In the process it promised an excess of everything to both residents and visitors alike. But now, with the fossil ground water all but completely depleted, the Colorado River yield was being shared by two other water needy/greedy states, and limits to growth loomed.

In short, what had once been magic had now become maddening, beginning with the searing heat being 10 degrees hotter in late summer than it was seven years ago, so that I wasn't even sure I had living, breathing neighbors at all until I saw their bags of garbage set out on the curb on trash days.

The traffic had gone from survivable to insane, with drivers swerving from inside lanes and crossing the highway at twice the speed limit, even in the construction zones that were everywhere. And the sound-wall along the interstate was decorated with a panel depicting the most famous of Vegas: Elvis, Marilyn, Sammy, and Frank—not one of whom I'd hold up for grandchildren to emulate. Graffiti "decorated" everything, even highway road signs so that if you didn't know where to turn off you wouldn't and couldn't.

The overwhelming sense was one of willful oblivion, human privilege living not in but with a desert, and on borrowed time.

Weather or Not

Weather is not something we have much of here in the desert. Day after day after day of relentless sunshine leaves weather reporters with little to do beyond seasonal warnings: Don't leave children locked in cars in the summer heat; secure patio furniture in spring and fall winds; don't drive through flooded intersections during rare rain.

So perhaps the weatherman was just bored that day that he went ballistic on the noon news. Reacting to the report of a neighborhood dispute over a golf course, he literally *lost it*, "What's the matter with you people?! You live in a desert! You're not supposed to have lawns and golf courses. If you don't like that fact, go back where you came from!"

This character was known for his outrageous outbursts, so probably no one was paying him much mind. But I was new in town, and stood in front of the T.V. transfixed. I listened.

And I heard him. He was right to be concerned about changing the landscape to imitate the Mid-west and Northeast. Already the constant irrigation for growing non-native plants had brought mosquitoes, humidity and allergies into the desert.

Human refusal to live within natural limits is taking its toll: the fossil water aquifer that created "the meadows" (Las Vegas) in the first place is all but gone, drained to supply golf courses, lawn carpets and trees, and casinos with dancing fountains.

While scrambling to increase our allotment of the Colorado River and trying to grab water from rural aquifers up state, the water authority also offered rebates to homeowners who replaced their lawns with native plants, stones and sand.

My home-owners' association, appalled by the increasing cost of watering the grass in our communally owned backyards, debated the option of replacing grass with "unthirsty" plantings.

Folks not before seen at HOA meetings showed up to rail against the idea, claiming they had bought their unit because it had grass and trees like back home. Never mind that most of us came here to escape the winter weather elsewhere, the very weather that's needed for lots of trees and luscious lawns. Tempers flared, neighbor turned on neighbor, and retired seniors became preschoolers who hadn't yet learned how to get along with others. Intimidated, I rose to speak anyway, suggesting that removing the grass was the "right thing to do."

Which of course it is, for, whether or not we want to accept the limitations of living in this "land of little rain," the desert will have the last word. It is just a matter of time.

Meantime, we did remove the grass. "I could hear each blade screaming," one neighbor whined.

I won't get into what happened when we had to take down the pines the developer had put in place for "curb appeal" that had started cracking retaining walls, lifting up sidewalks.

For I have become a "suspicious person," someone for the community's emergency response team to keep an eye on: their list conflates terrorists and environmentalists, marking me as a potential eco-terrorist. My driving a Prius, and my partner an electric car, makes us guilty. But it's the right thing to be!

The Truth of Light

If I become a westerner, it will be due to the light. Here at the edge of an archetypal American desert, dawns and dusks once painted such stunning streaks of pink across the wide horizon and inside my rented apartment that I found myself walking into walls, totally disoriented.

So when I returned here to settle and retire, I intuitively knew what I was looking for. As I prepared to purchase my duplex in Sun City Summerlin, I just assumed it had a kitchen as I went from window to window, checking out the light.

With *my* two hundred and ninety degree panorama, I plot the sun's yearly movement across the eastern horizon, watch the full moon rise on some nights, see the mountains emerge from the predawn darkness like consciousness at the beginning of human time. The afternoon light that pours through the clearstory windows fades into night's sprinkle of stars. And after rare rainstorms, I watch rainbow arches bless the valley.

At daybreak I focus on a lit candle as a way to mirror the glow that is growing and throwing itself into the new day.

At day's end the process reverses, and the light goes out with its glow slowly shrinking back into a flickering candle, but not before it sets off soft pink shades in stucco and stones, turns the mountains from sun burnt orange to shadowed blue.

This ritual of lighting a candle at daybreak and dusk is to participate in those grace-filled gateways to amazement, and open to an experience of at-one-ment with our planetary movement towards and away from our Star.

I know we collectively call these times Sunrise and Sunset. But the sun doesn't rise and set: we do.

We've known that truth since Copernicus, Kepler, Galileo, yet we still use our old names for that time at each end of each day when we see and now know that the world is infused with the Light just now reaching us from the beginning of Time.

No wonder the ancient ones worshipped the Sun. No wonder with Jesus the Sun became the Son. No wonder Hopi mothers still present their newborns to the dawn for a blessing.

Quantum theory suggests that all stable matter rests on an underlying sea of light. Light is everyone's Ultimate Reality!!

Lunar Eclipse/ Cosmic Christmas

Much of what we do over the winter solstice reflects a fascination with the cosmos. Take, for instance, the popularity of the poinsettia plants that appear in stores for only this season. Brought from Mexico by our first ambassador there, Dr. Joel Poinsett (I've visited his historic home in Charleston, S.C.), its blossoms delight us because they're the shape of the stars.

And of course the lit Christmas tree that has become the centerpiece in many of our homes originated with Martin Luther when, as the story goes, he was inspired by the sight of tall evergreens against the starry sky. So he cut a fir, brought it home, and placed lighted candles on its branches for his family.

Plus there's the image of Santa's ride against the night sky: an incarnation of a Scandinavian god who rode through the world at midwinter, bringing reward or punishment.

Perhaps we pay attention to the starry sky of this season because there are more hours of darkness, and the fallen leaves have opened up great swaths of empty space.

For as Lewis Mumford suggests: probably we humans were sky-conscious and season-conscious long before we were self-conscious. Thus, in this season of shortest days and longest nights, perhaps something in our DNA connects us back to our ancestors who shivered in the cold and darkness for far longer in human time than we've been cozy and warm during our long winter nap.

Maybe our dreams of a white Christmas are memories of that reflective *aha* during the last Ice Age, "When something within humans turned back on itself and took an infinite leap forward," as Teilhard de Chardin put it.

In that threshold moment, all human potential became possibility, from crops to civilization to spacecraft.

With that in mind and in honor of the season, I bundled up to sit outside in the cold darkness of a predawn in order to watch a lunar eclipse. As my mug of coffee kicked in, I suddenly remembered that I have a spotting scope. Looking through it called forth a *wow*! that I couldn't suppress, even in my concern over awakening sleeping neighbors. *Wow* indeed!!

And I thought of Galileo turning his spy glass from enemy ships on the horizon to the lights dotting the heavens: how he confirmed Aristarchus's observation, made back in 280 B.C., that the sun, not the earth, is the center of our planetary system.

Aristarchus had figured that out while watching a lunar eclipse! From the size of the earth's shadow on the moon, he deduced that the sun had to be much larger than the earth, and, therefore, it made no sense for the sun to revolve around a smaller body. Thus he put the sun in the center, with the planets in orbiting around it.

We've been trying to live into that realization ever since! We still say the sun rises and sets, while knowing full well that WE are doing the moving. Perhaps it is easier at the winter solstice to grasp the notion, the motion of our earth orbiting its star, especially if/when we can take time to stand still, stand still in the starry darkness and make a Christmas wish.

For long before this season became about enhancing retailers' bottom line, or giving a "vitamin shot to Wall Street," and even before it was about the birth of the Son, the winter solstice celebrated the wonder of the our star, the Sun.

From Angels to Astronauts

How utterly fitting that it was on a Christmas Eve that I walked away from ministry and out into the starry silent night.

After fifteen winter solstices of trying to speak meaningfully to humanists and Christians, agnostics and Universalists, pagans and Unitarians, atheists and mystics (while children secretly listened for sightings of Santa being tracked and broadcast by NORAD), my final sermon was a message not of angels but from astronauts.

Exactly forty years after that Christmas Eve when three astronauts orbiting the moon read to the world from Genesis, I preached on the new-born gift of that evening: that glorious view of our planet ascending into heaven, taking its rightful place among the stars. The Big Bang incarnate appeared as a baby blue sphere swaddled by clouds, lying in a black backdrop of infinite space. In that sacred moment, everything shifted: Heaven became the heavens, and snowflakes became star stuff. And wondering turned into wonder as the humans who had trained to explore the moon ended up discovering Earth.

Fueled by Cold War fear and scouting out a future landing sight on the moon, the space race was intercepted by surprise: "Oh, my God, look at that picture over there. Here's the Earth coming up. Wow, that's pretty. You got a color film, Jim?"[1]

1 from NASA's files

Suddenly, our geopolitical globe became everyone's Earth.

While separately tending our own humble lives, we each received the Good News simultaneously. Astronauts became magi, erstwhile prophets proclaiming a whole new world-view.

Yet, in our era as in all previous ones, we the people have refused to heed what our prophets keep trying to tell us.

Coming back from her final space mission, concerned Shuttle Commander Sheila Collins couldn't get T.V. coverage: ABC's lead story that morning was on the danger of driving in flip flops, not that of receding glaciers and expanding deserts.

So that lo, these many years later, NASA's James Hanson must get himself arrested to warn that our fossil fuel burning human misbehavior is about to be game over for the planet.

But he is not *out there* alone. Now the image from that long ago winter solstice has become part of the collective psyche, embedded in our conscious and unconscious Mindset. Hundreds of thousands, if not millions, of Earth's disciples are rallying around the planet, demanding that our elected leaders lead in the effort to contain the oncoming climate catastrophe.

Those of us alive on that Christmas Eve were brushed by the angel wings of our better human nature; whether this dawned right away or came upon us years later, a brand new consciousness and conscientiousness came forth from out of the stars on that silent night.

Leaving ministry was my way of saying *yes* to doing whatever was mine to do for the sake of this vision of Earth.

The Measure of Things

Whenever I get too full of myself I do this: Using my body to measure our planet's time within the 14 billion years since Time began, I stand up; my feet planted on the ground mark the beginnings of earth at the birth of our solar system 4.5 billion years ago; the first living cell appears at my ankle (3.8 bya); life's common ancestor is at calf level (3.5 bya); multiple cellular life emerges at my knee (3 bya); DNA exchange comes into being at hip level (2 bya); plants and oxygen come about at shoulder height (1 bya).

Then, raising up my arms: The largest explosion of life is at my elbow (500 mya); dinosaurs come and go just below my wrist (70 mya); at my wrist's bone is when humans appeared (2 mya). In the span of Time between the two lines wringing my wrist, modern humans walked out of Africa and landed on the Moon.

The 500 million more years that our planet may support Life can be plotted from wrist to fingertip. Sitting down with my hands open upon my lap, I see my personal lifeline etched on one, across the other is the whole arc of evolution. Yet the continued viability of that narrative now lies in human hands.

To grasp my place in this Age of Human Hubris, when we truly have become the measure of things, I go out into the desert and press my palms down on the sifting, shifting sand.

Chewing Sand

How delicious to be an elder...forbidden to get dirty as a child...now getting to play in the sand during paleontology digs!

Because Nevada Friends of Paleontology always met on Sundays, when I retired I was freed to join them in digs around an ancient watering hole out near the Upper Las Vegas Wash.

This fossil rich site was accidently *discovered* when contractors planting power lines kept tossing bones out of the way, and someone finally took a close look at them: remains spanning 200,000 years included wooly mammoths, camels, ground sloths, horses, bison, and giant North American lions. Now there's a move to preserve this whole area by creating a National Ice Age Park, which was my first clue that it was here.

Tromping out to the site too soon after a spring rain, we find water filling up the pit. Steve climbs down into it and begins bailing with his sand bucket; I join him with mine, replacing his full one with my empty one. Soon there's a whole reverse bucket brigade in progress, and before long we've displaced enough water so that Aubrey can set out the grid.

We each claim a spot to sit upon, begin scraping away, taking down the little hill that still covers the layer we're after.

A plop of congealed mud quivers in my hand, and I suddenly realize how it may have happened: how some early ancestor likewise held and beheld a well-defined blob of wet sand, noticed how strangely it stuck together, and began to fantasize, *what if what if...?*

I show the mud plop to another amateur digger; she offers to take it home and fire it in her kiln. As it hardens it will turn a psychedelic shade of green she'll tell us about later.

On another, dryer Sunday it takes three hours of moving dirt before we uncover a mammoth tooth the size of my fist.

Time stands still while Josh brushes it off, fixes it with fingernail polish so it won't deteriorate, then lifts the beige hunk of enamel up out of the last Ice Age and into Today.

We amateur and professional paleontologists alike stand around the edge of the quarry, wide-eyed with amazement: mammal contemplating mammal, we welcome our extinct ancestor into our era of Time.

We stare in silence at the remnant from a skull that once enclosed a brain from which evolved our own ability to bond, a brain that contained the potential for our human compassion.

My mammalian brain wants to go off in two different directions simultaneously: forward into the other human layers, those of rational thinking and intuitive knowing, and backwards on down through the layers of the personal consciousness and unconsciousness, to the collective unconscious, to the family, to the clan, to the large group, to the primordial ancestors, to the animal ancestors, and finally on down to the core Fire at the ground of all Being, and on out into the cosmos, and the star stuff we all come from.

That I can do so is the real Mystery here, that I am an EYE. So it's no surprise that I'm disoriented while driving home. Back in today with a honking black monster barreling down upon me like the coming climate catastrophe, belching carbon sequestered by life forms from even earlier than those just under this pavement, I grieve that we people have become so entombed in our wombs of metal that earth is losing its truth claim upon us.

I would if I could command everyone to spend some deep time digging in dirt.

Stripping off my sand encrusted clothes and dropping them onto the laundry room floor, (Mother would be horrified), I sprint through the house to the shower.

Life is meant to be messy, I tell her in absentia.

The spraying water creates a mudpack facial as I rinse out my hair, pastes powder-fine sand into my skin, cakes my pores. And although it'll all wash away, I will be forever chewing sand.

Windy

The essence of the desert is wind. Ed Abbey said so. But reading about it in *Desert Solitaire* is one thing, experiencing it first hand is quite another.

Gusts up to 50 mph were forecasted to become higher when we questioned our camping plans. But the campground was only fifty miles away, so we could always return home.

Four retirees, we had the luxury of an early arrival and a choice of camping sites. We selected one that seemed sheltered from the oncoming wind, and spent the afternoon poking around in the fiery red sandstone bluffs that gave the state park its name and its fame.

And yes, it was windy: we didn't dare light the campfire, and had to secure supper with lots of on-hand rocks.

We were to be two to a tent: one was set up by sunset, but we gave up on the other one soon after it turned dark.

Huddling between collapsed layers of canvas, I kept the mesh patch of a window over my face so I could breathe, as well as to see the stars as they moved across the firmament.

Staying awake to watch the glorious cosmic show was no problem, thanks to the wind. To say that it roared would be an understatement: it was more like the sound of a train coming out of a tunnel in an explosion of noise, pushing air ahead of itself. All you could do was to brace for the onrush.

There was to be precious little sleeping. All night long bursts of wind were interspersed with the sound of other campers driving away.

Choosing to stay, we could just lie awake, wired in full alert. I had visions of being blown with the sand across the Southwest, ending up in a solidified dune somewhere in Utah, Arizona, New Mexico, or/and Colorado. What story would my fossilized remains conjure up for future paleontologists? Would they know that the two breaks in my elbow came from skating: the first on the sidewalk at age five, the second on the ice at age twenty-five? And how could they tell I had relished my last moments of consciousness by living into my nickname?

"Why do you call me Windy?" I ventured to ask my high school boyfriend's father. For from the moment we met he'd called me that, and would do so well into my adulthood.

My mother had wanted to name me Wendy, but worried that classmates would dub me Windy Wendy, so named me "Gail" instead. And now I was being called "Windy" anyway.

"Because a gale is a wind," he said simply. "Is that it?" I pressed, half afraid that the real reason was because I talked too much, though actually only to him. A painfully shy teen, in spite of being head majorette as a high school senior, I kept my inner self carefully protected, except in conversations with him.

And he honored that part of me no one else cared or dared to see, myself included. His observation: "You love humanity, Windy, but you are not so sure about people…." haunts me yet. My "windy" self gravitated towards flying planes and writing poems even while I nursed, taught, mothered, and ministered. Now lying here out in the wind, the very essence of the West, I hear his laughter.

By dawn I am steeped in gratitude, not that the wind has stopped (it hasn't), but that somebody knew my true name.

Look!

Northeast of Las Vegas, in the stretch of spectacular scenery that's been preserved as a state park, one popular trail winds through a red sandstone canyon of petroglyphs.

Arriving early in the day, before the crowds, I find water unexpectedly flooding the entrance. No problem. It helps to think like a desert bighorn sheep and simply sprint up and over the slickrock, picking up the trail beyond the water. No wonder, then, that bighorn sheep petroglyphs catch my eye first: a line of animals ascend the fiery red rock. Why were they pecked into the black varnish on this rock face? No one seems to know for certain, but there is endless speculation.

I don't think much more about it until later. While I'm leaving the park the back byway, a bighorn sheep crosses the road in front of me. Awed, I stop to watch it; it sprints up a hillside where several sheep are lined up along the ridge. A few more sheep are slowly coming across the road, single file.

Then I hear the sound of motorcycles roaring our way. The first rider notices me in time to stop…to see what I'm looking at. A second rider does likewise. Soon we are quite a group, all of us standing and staring in shared silence as the desert bighorn sheep make their way to where we can't drive.

Perhaps this was the point of the petroglyphs: Look!

Look! As in the *Dick and Jane* readers of my childhood, the need to share our wondering seems to be a primal human impulse! As we slowly disperse, the bighorn sheep, the bikers, and me, I vow to continue the process.

So here now I have: Look!

Wild Horses and Burros

"Do Not Feed Wild Horses and Burros $500 Fine" Five years after I left Nevada, the state's quarter was minted. It depicted wild mustangs racing the wind over a sage dotted landscape, a scene I had actually witnessed! The sight had sparked my imagination in a way I'd not experienced before that brought Time forward, perhaps because horses are native to this continent. But some 12,000 years ago they migrated over the land bridge onto the Asian steppes, not to return until Europeans brought them back in their new world bound ships.

I have yet to see them again in the five years since I've returned, except in old cowboy movies and new documentaries about wild herds being secretly slaughtered to feed a growing international appetite for horsemeat.

Instead, I settle for spotting the non-native wild burros roaming the canyon lands beyond the city. Abandoned by prospectors when gold and silver mines played out, these horsey relatives now disrupt habitats for native species. Feral burros look so tame that tourists feed them from cars and turn them into a nuisance. So it should not have been such a shock when I saw a burro plodding along the highway in the bike lane.

Still, I enjoy coming upon a family of burros out on First Creek Trail, after stepping over their dung piles for miles. Like the tourists, I too am intrigued. I've even sent out cards that showed them gathered behind a barbed wire fence upon which was pinned a Christmas wreath complete with a red bow.

I delight in keeping my distance, watching them watching me, with our mammalian bodies breathing in the same rhythm.

They seem as solid as common sense, as grounded as a good idea. But something seems to be missing, something like imagination when it runs wild and free and far beyond what is immediately needed. With the loss of wild nature, what also gets lost in human nature? Perhaps so many of the big issues of our era seem unsolvable because of diminished imagination.

Revelation

How I loved getting lost, setting off over slickrock into the silence of solitude and the sweet scent of sage, sandaled feet bathed by silk-sensuous sand. But something was missing: the sign at the trailhead claimed it led to water, as most paths through the desert eventually do... but I kept getting stopped by the outcrop of building high boulders backed up in the wash, and could not find any way up, over, under, around or through.

When curiosity finally trumped the need for solitude, I said yes to a friend's suggestion that we hike together to the Calico Tanks. Arriving where I'd been stymied, she simply sprinted up a ledge barely the width of a hiking boot, and kept on going.

I followed, throwing my weight forward in the hope that shear momentum would overcome the pull of gravity and keep me from falling backwards down the steep incline.

Soon a whole other world spread out before me, as if I'd come into a sacred place through a hidden gate. Stairs appeared as though out of nowhere: huge blocks of rock led up to a plateau. I would have been content to stop right there, sit still to merge with the sky blue, rock red, and Manzanita green.

But my friend kept on going, heading steadily upwards, following a trail that kept disappearing as it wound over rock.

I had to trust her guidance. It felt like a spiritual act, as if I were following a Moses or Jesus or Mohammed or the Buddha. But when I saw the petroglyph Sun carved into the black patina of the redrock, I knew that out here the sacred Being is Nature.

And my Transcendentalist sensibility also knew that any spiritual transformation would not come about as a once and forever conversion, but as a call for continuous cultivation.

Thus I paid strict attention to the rest of the route, so I might return and resume the inner journey this place invited.

I never came back alone. This was not just because it is recommended to always have a hiking companion for safety, which is good advice as this same friend broke her toe on this trail during another trip, and I talked her back down to the car.

I don't come alone because I need to share the spiritual excitement of getting lost, seeking the path, and finally finding a footprint (though not always or only of a human) to follow...then pushing ever onward beyond all ones accepted limitations.

When the trail ends, another decision presents itself: climb down to the seasonal basin, touch in with its wetness, then spread out a communion of trail mix and brought water.

Or, continue straight up the side of the canyon, inch along its wall while ignoring the five hundred foot drop to the pool, and crest the summit that offers a polluted view of Las Vegas.

In the several times I've been out here, the path has never been the same. I'm still not quite sure that I can find the way, especially when detoured by pockets of water flooding the trail. Of course the time of day makes a difference, for the sunlight and shadows rearrange everything...so that nothing's familiar. Or boring: It's always a physical challenge, a spiritual adventure.

For up there I can stand on a literal line in the sand, step back and forth across colors that create the Calico tapestry for which this whole basin has been named. This line reveals the Earth's texture as a unitary Text, and Nature's script as a universal Scripture…a Revelation that could unify us all.

Perhaps next time you'll come too.

Tablet of Stone

I sit in the back of the class of fifth graders. We are all on the floor, listening to Mr. S explain the topography of the Grand Canyon. We have recently spent two days there on a field trip; I went along as writer in residence.

Now it is after lunch, and, like me, many of the children (the other introverts?) who need a nap struggle to keep from nodding off. It is at some point in this mental stupor that it comes to me: it does not matter one whit what Mr. S teaches these children about any topic. What they will always remember is his kindness, his concern for each of them, his firmness in demanding the best each child has to offer, his absolute dedication to his profession and all that it means to activate and influence these young minds. That he is also a gay man in a partnered relationship is irrelevant to anyone in this room.

After all, the kids have met his partner, and his dog, and are fond of both. It is simply not an issue. It is also no longer an issue in the state of Nevada. Just yesterday, before leaving for a hike out in the Calville Wash, I examined the framed certificate of Domestic Partnership on display in their shared home. At some point I will officiate at their formal commitment ceremony, and I am delighted for both of them. So is the partner's Latino family. Not so much Mr. S's family of origin: they are Mormons and their scripture defines this relationship as an abomination. While they too adore his partner, the fact that Nevada has issued a certificate of partnership sets up a conflict between two important authorities: sacred and secular.

It is a challenge being the different, difficult outlier in any situation, especially the family. So my heart hurts for this beloved teacher and his partner. We unpack his family's reaction on our long hike through earth's layers of deep time.

Tall and strapping strong, Mr. S carries a backpack containing the extra water, the lunch he has prepared, fruit, the sweet and salty snacks, plus the hiking poles for his partner who is busily taking pictures from a tripod.

He hefts the dog over strewn boulders as he shares how his brothers congratulated them upon the announcement and seemed to be happy about the legalized partnership. The sisters simply left the room, one sobbing.

I suggest this reflects the difference between male and female brain structure and socialization: men can and do compartmentalize the many aspects of their lives: family goes here, work over there, and any religious beliefs go elsewhere. Women, on the other hand, perceive the world as a whole, with everything interconnected: if you change a perception in one area, it affects all the other areas simultaneously, and sets up a disorienting disconnect until you can re-integrate all the parts.

The slot canyon walls loom high above us, the sliver of sky is darkening with clouds and we definitely do not want to get caught in this wash during a flash flood. Trees smashed to sticks caught in the crevices attest to the power of the water.

Hiking back to the jeep, these two men who epitomize the sanctity of personal integrity take turns helping me haul out a slab of pale green rock. And suddenly my heart hurts more for anyone who has compromised personal truth in order to please and appease the expectations of others, everyone who has denied or defied who they really are and what they are meant to be and do. They are the truly Lost Souls.

Stumbling a bit under the weight of the tablet, we joke about what new commandment might be inscribed upon the flat surface of this stone. What emergent wisdom might carry humans forward rather than keep us all mired in the past?

Perhaps we would take it into school, and ask the children.

What Happens in Vegas

Boarding my flight back to Las Vegas, I absentmindedly try to stuff my all weather coat into the overhead bin. A voice from the seat behind mine lashes out, "Watch what you're doing! Can't you see there's a wedding gown up there!?"

Stunned, I stop. And look. Indeed: a bulky white garment bag is puffed out across space meant to be shared by several passengers. Ah, is this the archetypal Vegas Wedding couple?

As the groom-to-be settles in with a series of drinks, his behavior goes from annoying to obnoxious to belligerent as we wend our way across the continental United States, confirming why I repeatedly refuse to perform weddings for couples coming into Vegas from elsewhere. I believe it is totally unethical of me, as ordained clergy, to join two people in marriage without first helping them examine their family histories and personal character traits and then laying out the issues that might trip them up along the unfolding road ahead.

Now a full-blown drama is playing out behind me. When the passenger beside me unwittingly pushes her seat back, the male half of the wedding couple begins relentlessly kicking her seat. This goes on for the full six hours, The female half of the couple is cowering against the window, perhaps seeing her situation in a whole new light.

Who is this man she's about to marry? Does she sense she is trapped? Is she thinking she can change him with her love? Perhaps it's simply the reputation of Sin City that brings on such bad behavior....

As we get up to exit the plane, I look over my shoulder and send a silent blessing to the bride while the groom wrestles her wedding gown out of the overhead bin.

"May our chamber of commerce's promise come true for you, 'What happens in Vegas, stays in Vegas.'" Amen.

Oasis of Sensibility

The meadows (*las vegas*) are celebrated in *Miracle in the Mojave*, a short film narrated by Martin Sheen and shown daily at the Springs Preserve. Created when ancient water sprouted up through a fault line, this riparian oasis attracted life, mostly migratory: birds and ice age mammals and pre-historic peoples.

But when we the people decided to stay put instead of pass through, we repeated the story of all previous civilizations who overwhelmed their natural resources: within half a century we've depleted the aquifer that took millions of years to fill up.

Yet we are the open-ended species; through us instincts became intelligence, reflexes become reflection. We can learn new responses, unlike the demonstration falcon that has just left its trainer's arm and headed for the artificial aeries of the Strip. I watch it fly off, leaving disappointed children behind.

The child I once was wanders the sustainability exhibits, appalled at the adult hypocrisy of, "Do as I say, not as I do." The elder I'm becoming fights back the cynicism of clearly knowing what we need to do, but that we don't and/or won't.

So I visit this oasis of common sense to witness what IS doable, from the desert demonstration garden to the solar paneled parking lot. Even the award winning café is a salute to sensibility: the artfully prepared food is served on glass, not plastic, plates, beverages in compostable cups. Now if I could just forget that triumphant staff tweet from the new Speaker's office when Congress last changed Party hands ("Plasticware is back!") that reversed all previous in-House green initiatives...

Carefully scooping half of my lunch into a biodegradable container, I can almost dare hope that Thomas Berry was right: we're transitioning out of the Cenozoic and into the Ecozoic, an era when we humans become mutually beneficial to our Earth. But for now we seem to be trapped in the Anthropocene Age, with humans poised to create the next great extinction.

Children's Discovery Trail

It's not a walk in the park, that crocked trail that winds through towering walls of timeless stone, requiring all hikers to scramble over boulders amid Manzanita trees that scratch any exposed skin in a thicket of willow alongside a trickle of water.

The final few feet squeeze you through a narrow passage to be carefully navigated by hands pressed against boulders.

Only then can you come forth into a clearing that presents as a Haiku poem:

Seasonal water
Falling down rock face stitching
Blue sky to green earth.

But before I can fully live into this Zen-like moment that merges the now here with the nowhere, the towering silence is shattered by waves of children's voices exclaiming "WOW!"

I'd brought my grown up daughter out here when she visited me in Vegas. She kept looking for ocean to go with all our sand, but she was eons too late. All that's left is ocean become rain become snowmelt now sprouting from cliff side before going underground and heading for what remains of a Colorado River system that no longer reaches the sea.

Now school children burst into the space far ahead of toddlers clinging to grandparents' hands and babies carried by parents. All arrive as if going to church on this Sunday morning.

Out in the natural world instead of inside a shopping mall, away from their computers and minus their cell phones, humanity's hope splashes in the pooled water and climbs on the rocks, fully engrossed in the no where become now here.

Tortoise Nephew

"Aunt Gail is here," my friend announces as I follow her through the patio door and out into her back yard.

I crouch down and hold out the rose I've brought from my yard as Nevada's state reptile turns and heads towards me. Tad is a handsome desert tortoise: the markings on his shell provide proof positive that he's not just a walking hubcap. He extends his beaked head and reaches for the flower; I guard my fingers from his chomp. He's hungry after his winter brumation in the burrow built just for him in the middle of the yard. Tad has been part of my friend's family since he hatched 20 years ago, the offspring of a pet pair of adopted tortoises.

Watching him slowly crawl around the yard and inside the house conflates the time between now and when dinosaurs roamed the earth. The current version of this species has been around for 2 million years, and is now threatened by humans.

Development, ORV traffic, grazing, and shooting have contributed its being listed under the Endangered Species Act, then moved to the emergency list five years later because of URDS, a fatal respiratory disease exacerbated by the stress of human impact on their environment. Then the night before the federal law went into effect, undaunted developers sent out crews to bulldoze the tortoises so far under they would never be found, to avoid the delay and expense of removing them.

I now live in what was once prime desert tortoise habitat. I've gone along for the ride when my friend put Tad into a box on the back seat of her car and drove him around hoping to frighten him into peeing and pooping. And I have accompanied her on snooper expeditions when, as a volunteer with the Tortoise Group, she's been called out to check up on reptiles reported late coming out of their burrows.

Lying on our stomachs on the ground in front of the opening, we'd guide the long tube with video camera into the burrow, probing for signs of life, worried owners standing by.

Like these owners, local school children eagerly await the re-emergence of tortoises. There's even a contest to guess the date when Mojave Max will wake up. Depending upon warmth and light, this event marks the unofficial beginning of spring.

We're always glad when Tad rejoins us for a new season and celebrates spring by gorging himself on the dandelions that my friend carefully cultivates for him each year.

It is almost as if non-human species seem to know how and when to garner allies in their fight for survival.

One case in point is a woman whose memorial service I was asked to conduct. Initially trained as a wildlife biologist, she'd moved to Vegas at midlife to pursue a Master's degree in zoology, and chose the desert tortoise as her thesis topic, "Because of the dearth of information." Her field studies were some of the first, and are still widely used. For thirty years she was devoted to the study, protection, and recovery of the desert tortoise, and contributed to several Country Conservation Plans, as well as cofounded the Tortoise Group.

The service was packed with a *choir* of advocates. The eulogy was delivered by the well-known tortoise activist I had read about years earlier in a book on the Mojave Desert.

The service felt like an initiation into the extended family of the Desert Tortoise, and I'm humbled that Tad has adopted me.

Meteor Shower Meet-Up

With August's promise of a meteor show, three things are needed: clear weather, open sky, and stamina to stay awake.

When conditions are just right, I head out beyond the sky shine of the Las Vegas Strip and into the deepening darkness, expecting solitude and silence in the national conservation area but instead find a mess of cars lining the rural highway.

Finally pulling into a little used parking area just outside the public park that closed at dusk, I find it is filled with people. Vehicles line up along the fence, all pointing toward the barely visible shape of the escarpment on the northwest horizon.

It feels like a veritable block party, with the shadows of adults sitting in folding chairs, children laying about on spread blankets, and teens lounging on car hoods and rooftops, as if it is two in the afternoon rather than two in the morning!

Oohs and Aaahs greet the sight of any streak of light. Those who missed it are assisted by those who caught it, which is helpful as I had no clue as to what I was looking for.

Excitement builds as more of us see what we made this post-midnight pilgrimage to witness. Soon star stuff is celebrating star stuff, co-creating a worship service that could rival Christmas and Easter, one that is altogether fitting for the new view of the cosmos in which we all live and have our being.

"Wow!! Did everyone see that last one?!"

Sav(or)ing Red Rock

It is usually in early mornings that the rising sun strikes the red streak of sandstone on the limestone grey mountain face and sets it off in a such a blaze of red-orange that it takes your breath away...and gives this piece of scenery its name.

But now it is late afternoon, and the winter sun is dropping down behind the escarpment, and all at once the whole mountain's infused with colors I have never before seen: multiple shades of muted red blending into grey: "pink" just doesn't name it. I am transfixed and wish I could pull off the busy byway to just be present in its Presence.

But I'm headed for a public hearing on behalf of this very view, so I continue winding along the rural road to the little village of Blue Diamond, some twenty miles west of Las Vegas.

In the deepening darkness I make only one wrong turn before finding the Quonset hut meeting place. Once inside, I am grateful for the winter jacket I rarely get to use here. It's cold and the locals are huddled under the heat vents at the back of the room. But as soon as the space fills up, the combined body heat thaws us all out, merges us into an unlikely alliance.

Hikers and bikers, climbers and cyclists, archeologists and astronomers, paleontologists and geologists, private citizens and public employees, all are seeking to voice their opposition for a proposed city on a hill just adjacent to our Red Rock park.

We're united against a developer who so completely embodies the archetype of greed that it's difficult to believe he is real…unless and until we find the latest newspaper report of his bankruptcy filings, divorce proceedings, or lawsuits.

His cadre of lawyers monitors public meetings, then come up with a new loophole to wiggle through while the community and county are busy dealing with a previous one.

This has been on-going for nearly a decade. After he purchased the nearby defunct gypsum mine, he filed with the county to construct a rural community; almost as soon as he received the permit, his plan "morphed" into a city of 14,000.

Now he wants the county to get him federal permission to run his construction trucks up through a swath of BLM land. That's why we are here tonight; the local commissioner wants to hear from the community. And she indeed gets an ear full!

Yet how can any of us ever adequately advocate for the sound of silence that would be shattered by construction, the scent of fresh air that would be compromised with thousands of daily car trips up and down the hill at build out, the sight of meteors streaking across the dark skies that would have too much night shine, the sense of deep solitude while touching in with marine fossils and wildflowers that would be paved over for a shopping center, or the taste of participating in the progression of life rather than mindlessly complying with the so-called "progress" that is plundering the planet for profit?

The developer knows how to play the game. After all, he helped write the rules, and bought politicians to pass them. But this time, he doesn't have enough commissioners' votes, so withdraws his request, and takes another tact: request a mining permit from the state. This ploy nearly made it without public comment, but someone caught on, and demanded a hearing. Another year, another December night of shivering in the Quonset hut, I sit listening through the smoke screen of words that mean the opposite, i.e. development as reclamation.

His permit is granted, a mere formality really. His trucks have been moving up and down the dirt road to his "mine" site through public land for over a year now. My partner and I have hiked a trail paralleling the road, counting them.

And we have hiked to the very edge of the BLM land, peered over at plateaus of scraped land: for more mining or for building foundations? My partner's curiosity takes him further, over yet another ridge three miles in from the scenic byway.

When I can bear it no longer, I drop back to find a crinoid encrusted slice of slickrock to sit with. This desert used to heal my heart, bring me clarity of thought, and soothe my spirit. Now this place breaks my heart, and fills me with a grief so deep I want to throw myself down on the ground and sob, then roll over kicking and screaming with frustration and fury.

Marine Fossils and Wild Flowers

For the late Elizabeth Tarbox, colleague and close friend.

While it is true there are few crocuses to scrutinize or daffodils with their death defying pollen poking up through snow, like those you wrote about back in New England, the Earth is renewing itself again this Spring…even, and especially, here in the desert...if you know how/where to look.

I would take your warm memory with me out to the fossil ridge where new Life is exploding through so many species of wildflowers we keep losing count: is that blossom #21 or #35?

Here we can celebrate the impulse towards rebirth and regeneration that has found its way into human stories from Persephone to Jesus. As we sit amidst remnants of marine forms that re-emerged after the Cambrian explosion went extinct, we'll wonder if these creatures knew that they too were to perish in the Permian extinction millions of years later?

Or is it only we humans who are aware that we are alive and will one day die? Your final year was all the more poignant when you let me walk with you through that shadowed valley. Grief and joy became utterly interchangeable as we laughed and cried simultaneously over that green blob of jelly escaping from my bowl and shaking itself silly on your dining room table the night before your diagnostic sentence of death.

Now from our platform on this fossil ridge, a shout of yellow and purple and pale green arrests our attention against the stony silence of antiquity's pink streaked brown mountains, reaffirming that only after flowers concentrated sunlight into seeds could our warm-blooded ancestors arrive on the scene, and Life take its fateful, faith-filled leap in awareness of Death.

And thus spring comes again, with its warning to expect death while we're not to be afraid to fully live. I try to, but Elizabeth, I don't know what to do with the grief that is growing stronger the longer I go on beyond you: for it yearly grows clearer that we humans are causing massive extinctions and that ecocide means genocide and will also become suicide.

It makes my astonishment at this year's crop of flowers almost too much to bear. I need your bravery to mediate my despair, your absent presence to help me pray for a miracle that might change human behavior.

Meanwhile, it is my behavior that is changing, and by spring of next year, when we both would have turned seventy, I may no longer be able to lie down and be as good as you were.

Wasteland or Wilderness

Perhaps it's a tribute to human ingenuity that early Euro-Americans could look at Nevada's arid vastness and see hidden treasures of silver and gold, gypsum and oil. However, there is one little glitch in our greed: once these resources run out, what's left in their place are ghost towns and waste tailings.

So it's no wonder that the next wave of attention paid to Nevada turned towards making the best use of its wasteland: our military relished the open empty space as a place for the testing out of its hardware, including the atomic bomb.

There's a pull out on the side of Mount Charleston a mere thirty miles north of Las Vegas where today you too can stand where tourists once watched atomic blasts for entertainment, some while sporting a special mushroom cloud hairdo.

And now, not far beyond the shimmering mirage of water held in the brown bowl of the desolate desert, a repository is up and running, ready to receive the Nation's nuclear waste.

Perhaps it's a tribute to the change in human awareness that brings many Nevadans out in protest of that project, and compels others to become advocates of this place by striving to protect it, and by legally preserving it as wilderness.

"In wilderness is the preservation of the world," wrote Thoreau in the era that launched our sense of Manifest Destiny.

The war he refused to support with his taxes, that act that landed him jail, was the one during which we wrangled the whole of the Southwest away from Mexico. We really only wanted more direct access to the riches of California, and had no idea what to do with the wilderness that lay in between.

The whole notion of letting it be for its own sake rather than for ours is a new one, born from the souls of first peoples, poets, hikers, spiritual hermits, and the like.

To count yourself among them would mean becoming an active and activated part of the earth's immune system.

It would mean aligning yourself with efforts to protect threatened species like the desert tortoise, the Mt. Charleston blue butterfly, and the Moapa dace.

It would mean agonizing over the irreparable damage done by off roaders and target shooters.

And it could mean wondering about the place of your own species within the bigger picture of things.

Wasteland or wilderness; exploit or experience; abuse or appreciate; useless or use less; the choice that is facing us plays out in the vast sea of sagebrush and sand that spreads between ranges of mountains that make up the land of Nevada.

Goddess Temple

It is not an apparition, that strange structure blooming from the desert floor some sixty miles northwest of Las Vegas.

At first I thought I was conjuring up that infamous image from Hiroshima, the haunting steel skeleton left standing after our nuclear bomb incinerated the city and its population, that dome that has become the centerpiece of its Peace Memorial.

This is not an inappropriate impression, as this temple of goddess spirituality was erected here precisely because this desert space is close to the Nevada Test Site, where atomic bombs were tested above and below ground for half a century.

The small stucco covered straw bale building is filled with goddesses from all of the world's major religions, and stands as a testimony to another perspective in an effort to balance the prevailing patriarchy's answer to everything: warfare's violence.

Mothers are the only gods in whom all the world believes Joseph Campbell claimed. And this makes psychological sense: all children come forth through women, but boys must learn to separate from the mother by making her other, while girls identify with her, as they will become mothers themselves.

Perhaps the human race needed both in order to evolve its fullest potential, but that balance has been tragically out of order for most of the past 5000 years. So that now humans have so violated the earth that we stand at the brink of extinction.

The largest statue within the temple is that of Madre del Mundo. Originally placed at the front gate to the test site and immediately confiscated by the federal government, this is a sculpture of a woman holding the world in her lap.

It became the inspiration for creating the goddess temple. Now this little oasis of feminine energy witnesses to another way of being in the world, with its goddess ceremonies, gift based economy, and dedication to peace.

With its four sides open to the weather, a fire pit in the center, and a ceiling that lets in the sky through a dome of seven interlocking copper hoops meant to suggest lotus petals, the temple stands as a symbol of healing in the midst of a radiation ravaged land. Overhead, fighter jets from a nearby air force base buzz by, and a lone drone slinks its silent way over the interstate, each a personification of Destruction and Death.

But then I remember that Nevada means "snow-clad".

And directly west, along the tops of the mountains that define the horizon line against the brilliant blue sky, a mantle of snow holds the nourishing feminine moisture that means Life.

Shoshone Sunrise

"Remove your hat," the Shoshone women admonished me as I joined the circle in the predawn cold, "Show some respect."

I was totally taken aback: cap, sunglasses, and water had become such standard equipment for desert living I was no longer even aware of them. Plus, I could have protested that in my childhood church, women were expected to cover their heads as a sign of respect. But that religion had not done well by these people now gathering here to welcome the sun….

Most of this landscape still legally belongs to the Western Shoshone: they never sold nor ceded their Nevada territory. They simply signed a treaty of friendship allowing pioneers to use an overland route to California and the railroad to be built.

But they badly misunderstood the way of white people: we have been trying to seize the land ever since, by right of eminent domain. Payment for it is being held in a trust by the U.S. government, as the tribe continually refuses to accept it.

Most recently, two Shoshone sisters have taken their case to the U.N and won a judgment against the U.S. for removing their cattle and horses when they refused to pay for grazing rights on the land that they've claimed for 10,000 years. But our president had already leased it to the gold mining industry.

Meanwhile, this landscape that is so sacred to the Shoshone has become the most bombed place in the world, with over 1,000 nuclear bombs tested upon it. In response, each spring the young men of the tribe circumambulate the periphery of their homeland, planting willows for healing.

And now I have unwittingly added insult to injury! I pull off my baseball cap and stuff it in the pocket of my jacket as we all join hands and begin to walk counterclockwise in a circle.

I am so overwhelmed with both personal and collective remorse that I can barely pay attention to where I am going. I'm unable to get into the rhythm of the drum with the placing of my feet until I suddenly notice a singular desert marigold sticking up from the path, its yellow face brightening with the sun's rays. In response, I pull my part of the circle outward, trying to make another path away from the flower.

The Shoshone woman who has criticized me does likewise, and soon we have the whole group carefully avoiding the fragile bloom. Once the sun has risen, and the full moon has set, the ceremony ends. I squat over the blossom to brush off the dust from the dance.

Everyone else has left, except for the woman; she's there still. As our eyes meet, our worldviews merge into our mutual respect for a singular desert wildflower.

Nevada Desert Experience
-Faithful Witness-

When I first heard about the Nevada Desert Experience nearly thirty years ago, I confess I didn't get it: why was an east coast friend wanting to walk across a desert protesting? While it was true we were involved in peace issues together, I firmly believed in working for change through existing systems.

I'm not so sure about that anymore. And while I'm still not sure about the annual spring rite of futility still happening in Nevada, since retiring here I am drawn to its Holy Week events, such as the Interfaith Seder at a campsite on the 65 mile route and the Easter Sunday worship at the entrance to the Test Site. And I am beginning to "get it," thanks to my friend Megan.

For recently, as Southern Nevada endured record setting summer temperatures (oh, but it's a dry heat!), our own Sister Megan Rice was experiencing the not so dry heat of a summer in Georgia, where she's in prison, awaiting her final sentencing.

I first learned about her arrest on Democracy Now. When Amy Goodman reported that an 82 year-old nun, along with two male Transform Plowshares Now activists, had broken into the Oak Ridge, TN, nuclear-weapons facility, I knew right away that it had to be Megan. And of course it was. How could it not be?

This frail looking white haired woman whose whole life has been steeped in simplicity and service, who occupied a single room in the Nevada Desert Experience compound and tooled around Las Vegas in an ancient Diesel automobile fueled by vegetable oil (also donated), has become a local legend.

In fact, she's been arrested so many times here in southern Nevada that she's part of the official oral history of the Nevada Test Site.

Her faithful witness to what the rest of us would prefer to ignore includes calling attention to the drone warfare being carried out from an airbase north of Vegas.

But her most persistent witness is on behalf of "the disharmony of our planet caused by the worst weapons in the history of mankind, which should not exist."

Originally charged with misdemeanor trespassing and looking at the possibility of one year of jail time, she has been convicted instead of felony terrorism that could mean up to 20 years in federal prison. The sentencing keeps being postponed.

Megan never expected to get as far the three of them did that night they hiked through the eastern Tennessee woods and the mile up to the ridge, guided by google and god.

They kept expecting to be stopped by guards; after all, this facility stores high level nuclear material that costs us tax payers millions of dollars a year to protect.

Yet these pacifists simply cut through three fences and walked right up to the building that is the nation's, "Fort Knox of uranium." They spray painted the building with Biblical phrases, and poured bottles of blood from an earlier resister, frozen upon his death so that he might participate on one last Plowshares action. They unfurled their TRANSFORM NOW banners and, with a small sledgehammer, began pounding on the base of the concrete guard tower at the northwest corner of the building...all while expecting to be arrested.

The lone guard who finally showed up had spent his life protecting nuclear facilities, and he knew the difference between a protester and a terrorist: he did not shoot to kill.

That act of common sense and moral courage cost him his position, his pension, and his health care plan, and left him with a sick wife, a foreclosed home, and the bitterness that comes from being made a scapegoat: he was the only one who suffered the consequences of a security system whose seven cameras had been out of service for six months.

Outraged government officials, appalled that such a high profile nuclear facility wasn't even "nun-proof," began holding hearings in Washington, D.C. Sister Megan was asked to stand up, then thanked for, "pointing out some of the problems in our security." Since then she's been locked up.

In a recent response to a letter I sent to her in prison, Megan wrote, "Have no concern for my comfort here. It is always a privilege experiencing and sharing life on the margins."

From now on whenever I walk through the desert around the NDE Peace Camp, where stones have been arranged in patterns that were meaningful to participants throughout the years, I'll salute the presence of a cloud of witnesses that now includes Sister Megan, and give thanks for their faithfulness. And I will carry her words in my heart: "The truth will heal us and heal our planet."

Leap Day in Las Vegas

Is this an extra day for playing catch-up with our Star, or is this to be a day for setting into motion the quantum leap of consciousness that is needed for a viable planetary future?

While we've been asleep, corporations and politicians, big banks and Wall Street, have been busy beneath the radar screen. And the mainline media is so obsessed with celebrity scandal (this week is focused on who wore what at the Academy Awards), that for any news we may need to be responsive and responsible American citizens, we must take our Secretary of State's advice and tune into Al-Jazeera World News. Stewart and Colbert on The Comedy Channel also tell the truth, if only to make light of it, making it easier to swallow.

But we're waking up now, and we're not laughing. There is so much gone wrong, so much to right, so much to do that we're scattered, frustrated, and often acting out among ourselves. Yet chaos means opportunity. And #F29 is the day to protest the American Legislative Exchange Council (ALEC).

Our gathered group is pathetically small in this city of two million, but here we are, come out into the delicious winter sun. Members of Occupy Las Vegas, Biodiversity, the Sierra Club, and the Moapa Paiutes, we are children, youth, students, middle agers, elders, retirees, the unemployed, underemployed, the homeless, home owners, snowbirds, seasoned and first time demonstrators all making this statement: we've connected the dots between our local energy monopoly whose coal fired power plant has been poisoning and polluting with impunity for half a century, and ALEC's rewriting of regulations that block EPA action on behalf of people and the planet, keep fossil energy profitable, ignore the changing climate. Such is the irony of this day: current sunlight blesses us as we disperse, determined to challenge the status quo.

Paiute Powwow

1. The Paiute Tribe has occupied what is now southern Nevada for over 2000 years; now they cling to a few pieces of land and try to survive. Never mind that one of Nevada's two statues in the U.S. Capitol honors Winnemucca (1844-1891), a Paiute who gave over 300 speeches defending tribal rights from encroaching whites, started a school where the children were taught in both English and their native language, and wrote her autobiography, the first book by a Native woman. Today her descendants host an annual Veterans' Day Powwow across from the Travel Plaza that funds the Moapa Paiute Reservation north of Las Vegas. I make it a point to attend because November is officially Native American month here.

The calendar says fall, but the weather still says summer, with temperatures soaring to brutal beneath the relentless sun. Indigenous families set up protective tarps for shade but the stands for visitors are exposed and uncomfortable as we rise for the Grand Entrance of participants parading into the arena behind American flag. But as a young Indian woman sings America the Beautiful instead of the unsingable Star Spangled Banner, the heat becomes inconsequential, and by the time a kaki clad Native veteran sings Taps, his plaintive rising and falling voice echoing off into the surrounding desert and to the horizon line of mountains, I am a pool not of sweat but of tears. I'll stay put through the variety of dances (fancy, jingle, traditional) to the Iron Man dance. For men only, this contest comes with strict rules: no stopping, no regalia parts falling off, and keeping up with the rhythm of the drums. Contestants start dropping out after twelve sets, the old men go first as this is clearly for the young, and two twenty something's make it through 17 dances before a winner is declared.

The grand prize is a thousand dollars...in food stamps.

2. When we next gather at the Powwow grounds, there are no stands or vendors or colorful regalia nor a Grand Entry, just the essence of the Powwow, which is empowerment. For the Moapa Paiutes are taking their power back from the energy company that supplies Vegas with electricity while poisoning their people with toxic ash that blows across their reservation.

When the aging coal-fired plant came up for renewal, the Paiutes decided enough was enough. Allied by Sierra Club, Earth Justice, Occupy Las Vegas, and Biodiversity for the final mile of their Earth Day march from the reservation to Vegas, the Paiutes have done more than just protest our local energy company. Rather their strategy became two-fold: shut down the dirty coal plant and construct a clean energy alternative. Working through the federal political system, the tribe secured permission and permits to construct an industrial size solar array upon their land. And when NV Energy refused to buy into the project, the tribe contracted it out to southern California.

We are but a handful of yellow (Beyond Coal) shirted people joining with the tribe on this new Earth Day. Together we march under the interstate and a mile into the desert, walking between railroad cinders and bursage bushes. Our goal is the place where the solar array will one day be; we arrive bearing paper panels to hold over our heads, because the construction can't begin until all of the desert tortoises have emerged from brumation and been relocated.

Before the picture is taken for the Sierra Club Beyond Coal website and the celebratory circle dance begins, the tribal president speaks to the people, "We were told that this could never be done, that this would never happen. Yet here we are!"

(NV Energy is now selling out to a non-local energy monopoly.)

Hobgoblin Inconsistencies

Surely that's not me in the photo of a human among the hobgoblins of Gold Butte. And that's certainly not me perched on a precarious slab of stone taking note of the surroundings, intoxicated by the expanse of contorted red sandstone that's totally inconsistent with anybody's preconceived expectations.

Maybe it is Emerson, sitting there beneath an endless sky, where shape-shifting clouds persistently, consistently rearrange the landscape. Is he listening as his intuition utters memorable lines about hobgoblins and consistency and little minds?

No, it can't be Emerson; I doubt he made it this far south on his trip across the continent to visit the great redwoods with John Muir. Although I really can picture him out here in this part of the boundless West that he believed we as Americans needed in order to stop being pseudo-Europeans.

So might it be a Paiute, someone who lived here long ago, back when first peoples traveling this trade route followed the red sand that had blown across most of the desert southwest?

Or perhaps it is Calvin, a Paiute of today, who comes here to pray for his tribe still living on the north arm of the manmade reservoir that has flooded so much of his homeland. No, it can't be Calvin; he's in the hospital with the asthma and a heart condition caused by fifty years of breathing toxic ash from the coal fired power plant adjacent to his reservation.

So then, who is that human being dwarfed by an arch of stone that is studded with petroglyphs riddled by bullet holes? And who do those hobgoblins seem to be glaring at….not me, as I scramble back down the rock face echoing Emerson's

"A foolish consistency is the hobgoblin of little minds, adored by little statesmen and philosophers and divines."

Sloan/Stone Canyon

After a teeth rattling jeep ride over five miles of dirt road running underneath power lines, you'll arrive at the mouth of a stone canyon, and park near a BLM sign that ought to read:

"Caution! Beyond this point you risk being haunted!"

A half-mile hike through a sandy wash gives you time to reconsider, then a barricade of boulders the size of small cars plugging the canyon creates another pause for pondering.

Is this remoteness a challenge that's meant to protect you from something....or is it protecting something from you? If your curiosity gets the best of you, and you scramble up over the boulders, you'll be surprised at how many people are already there. How did they get here; why have they come?

But soon your awareness shifts away from today, and all at once you are standing in the very place where ancient artists stood contemplating black stones large enough to...to...to do what, exactly? Did the sight of such surfaces spark creativity?

What if these volcanic rocks became aware of themselves by means of the petroglyphs that are now covering their surfaces. Through human hands, abstract murals emerged, compositions of wavy lines, circles, rectangles, "necklaces" of nesting curves punctuated with arcs and dots, all artfully arranged on individual boulder slates as if by pupils taking notes from an invisible instructor.

Stick figures and big horn sheep, feathers and fingers are not hidden away as in European caves; these are in plain sight!

So that now you too are standing back at the beginning of humanity's awakening to its potential, when the inner gifts of insight, imagination, and intuition were painstakingly chiseled onto rock, and communicated. It's as if Emerson's essay *Self Reliance* has been permanently set in stone! And while you cannot fully know what our prehistoric cousins were thinking, one thing is certain: you are bearing witness to the miracle of an individual brain tuning into the mind that would outlast it.

And what then of the haunting? Simply standing in Sloan Canyon, a site seriously threatened by human encroachment as Vegas sprawls southward into the Black Mountains, invites your discernment of where your life falls on the spectrum between creativity and destructiveness, inquisitiveness and irrationality, "transcendent aspiration and demonic compulsiveness," as Lewis Mumford describes the emergence of consciousness. For

"Man's own development and self discovery is part of a universal process that may be described as that precious part of the universe which has, through the invention of language, become aware of its own existence" (The Myth of the Machine).

The Great Unconformity

Between Lake Mead and Las Vegas lies a mountain of mysteries: Frenchman Mountain, tipped on its side as if a rug had been pulled out from under it, reveals a large gap in earth's geological narrative that is known as the Great Unconformity.

When I was first brought to this spot a decade ago, a local Boy Scout troop had just cleaned up the mess caused by years of shooting, dumping, and vandalizing, and the area had been newly, duly dedicated as BLM managed interpretive site.

Signage explained what I was looking at: two exposed earth layers separated by 1.2 billion years of missing history!

You can stand in the gap between them, touching 500 million year old limestone with one hand, and, with the other, literally touch in with our planet's 1.7 billion year ago basement granite and schist. The only other place you can do this is the inner gorge of the Grand Canyon. I know because the mule I rode from rim to river kept brushing me up against the same pink granite that's exposed here in southern Nevada. For, point of fact, this mountain migrated over from Arizona, and here you can, theoretically, simply stop at a pullout on the road to Lake Mead and see what otherwise would require a 14 mile roundtrip hike, an overnight mule ride, or a several day raft trip.

Yet walking the few hundred feet to reach this site can be more hazardous than any of the above, for it involves picking your way across a carpet of shattered glass, trying to recall the date of your last tetanus shot. And it helps to know the way, for all the signage has been destroyed over the last few years.

Still, I can't resist making a pilgrimage to this place on a regular basis. For some reason I need to sit in the great gap of the earth and ponder: transform, conform, deform, reform...all human created and related concepts.

But what are we to do with unconform, this buried erosional surface recording something absent?

Continual pinging from illegal target practice accompanies the rest of the climb up the trail, on through pink granite rocks that have been tagged with gang blue graffiti.

Cresting the ridge reveals all of the Las Vegas Valley, lately become the haven for bad human behavior that seems hell-bent on trashing the planet, beginning right here.

Okay, not just here: in my mind's eye I am hiking through conservation woodlands behind my brother's house in New Hampshire; he bends down to pick up beer cans carelessly tossed by an off-roader, shakes his head in disgust, and says, "The earth will shuck us off and start all over again, real soon!"

But being clergy, I protest: "No! Not yet, not yet."

For while people disappoint, humanity must be here for a purpose even while it appears that human consciousness and conscience have been eroded away by the floodwaters of consumerist greed, in a whirlwind of public corruption.

Thus is evolution's trajectory reversing, devolving us from human reflecting to mammalian relating to reptilian reacting? Will we too go the way of the dinosaurs? Or will we transform in time to save ourselves from the conditions we've created for our own extinction?

I hold onto hope as I straddle this Great Unconformity, a fragment of pink granite in one hand, in the other a slab of marine green shale, looking for a trilobite that'll answer me.

Keystone Thrust

How can it be that I've never hiked this trail until now? Surely I've known it was here all along, even as I explored the main part that meanders through lavender spring lupine. Yet this cut-off has mostly gone unnoticed, until now.

So why now?! There seems to be no rational reason, only the need to get away from T.V. and radio, computer and tablet, with their stress producing news of our day:

A recent rally on the National Mall with 40,000 people demanding leadership on climate change, beginning with the rejection of the dirty tars sands pipeline, has, as expected, gone largely unreported by the corporate media; what is unexpected is the discovery that, during that rally, the president was golfing with oil and gas executives in Florida. And this morning, the local stations were abuzz over the new oil exploration under way in Nevada as a boon for the economy.

This day is a cold overcast gray; the air feels raw enough to snow: perfect conditions for a shade-less trek. Nearly a mile in, we crest the ridge composed of grey Cambrian limestone and start downwards onto a plateau of red Jurassic sandstone.

The cause of this juxtaposition is contained in the name of this trail: the Keystone Thrust, a fault line that runs through this mountain range and extends into Canada.

The irony is not lost upon me: it is Canada's Keystone XL Pipeline that has become *the place* to finally take a stand for saying no to fossil fuels and yes to renewable energy sources.

Finding a secluded sanctuary, I climb up on a slab of red stone beside a lone Manzanita blooming in the frigid February morning, glad that here in the desert my birth month means early spring flowers instead of just late winter snows.

But cyclical seasons are not the main source of wisdom out here. Geologic time is.

And therein lies the frustration. The human mind may be able to measure but not grasp the Time that is opened before me: this fault line marks the point where the ocean once ended and this continent began. The layer of the earliest explosion of life (there are the horn shaped fossils embedded in the limestone) has thrust up and over the layer that, millions of years later, supported the plodding of dinosaurs and the scurrying of mammals that would one day become us.

In between these came the carboniferous period that utilized and then buried enough carbon dioxide to cool the atmosphere and allow for a whole new epoch of life. Yet it is this sequestered carbon we are now releasing back into the atmosphere that's reheating the climate. Were we to bring to the surface and burn all that's still out there to find, Life as we know it will be over forever. There are already mass die outs, and we too will go extinct. Thus we must leave the rest of the carboniferous age fossilized coal, gas, and oil in the ground.

Science has made this all quite clear for us. So why can't or won't we transform that knowledge into wisdom? Claims that it's inconvenient or obstructed by greed seem too easy.

Perhaps that question is what's brought me out here; though mostly I just want to bang my head against this immutable sandstone, that intractable limestone.

I sit close to the fault that drove one crustal plate down and under the other: if there's a key to moving from human understanding to action may it be here in this cataclysmic shift. But perhaps it is already coming to pass: that rally in D.C. might be the eruption of people-power on behalf of the planet.

Joshua Trees

The Joshua trees are blooming this spring! There must have been enough moisture since last year, when they didn't. So I must go out to a trail where I can get close enough to simply listen for what they might tell of their own story, instead of the one superimposed by early Mormon pioneers.

Travelers knew they were half way between Utah and California when they spotted these twisting denizens of the Mojave Desert, and named the species for the successor to Moses who prayed for divine directions to the Promised Land and received the reply to follow where the trees were pointing.

But these trees aren't pointing in any one direction! Rather, they appear to be lifting their flowers fist-like into the clear desert sky, as if in defiance of an upstart species that has scattered broken glass and plastic bottles in their habitat and is now threatening their existence by changing the climate.[1]

I find a rock that's flat enough for sitting where I can be with a large Joshua "colony"—a cluster of yucca like trunks. Some have arms that are budding, others display blossoms that look like lilies, while still others are clearly nursery

1. It is speculated that the explosion of this year's Joshua blossoms is due to two years of drought stressing them out.

plants under protection of a species that's been here for 200 million years.

It's message: it may be time to stop worrying about what we are making extinct and start nurturing what will outlast us. Stunned by the insight, I vow to celebrate Earth Hour this spring in honor of these Joshuas by powering down and lifting the human footprint from earth's neck to let it briefly breathe.

Raining Tree

In this season when the scent of honeysuckle, the sound of doves, or the sight of a yellowed leaf can break my heart, I pause with the planet as it prepares to tip away from our star.

In the pre-dawn on this longest day, I hike up to a neighborhood overlook to greet the sun, a commitment I keep with the cosmos. Watching rabbits nibble the grass, I try not to envy their seeming unconcern with a coming seasonal change

Sometimes I lament that as a human I am programmed to ponder, like the winter solstice Mary beholding her son: What new life did I say yes to bringing forth this year?...and what piece of the past still stabs at my heart...not as a regret, exactly, but more like a wistful *what if?* at certain crossroads, such as that long ago summer solstice I celebrated with my soul's twin by accepting the orange he offered, even though our lives were on different trajectories.

Now I ignore the hummingbird buzzing me until it rises like a phoenix in front of my eyes. OH!, of course: the dynamic of the sun's daily rising and setting, yearly dying and being reborn, is imprinted in our human psyche and transcribed into our sacred stories from the Phoenix to Jesus.

Even these rabbits symbolize something: the profile of a hare etched on the face of the full moon is a sign of temporal time, dying and being reborn into the eternal light of the sun.

Our human spirits know how to process these two dimensions of time: the now here and the nowhere become one and the same simply by paying attention, a mystical definition of prayer.

But what of this new sense of time unfolding for those of us who have lived through our own death and resurrection experience, that midlife passage from ego to transcendent self, and are now on the other side? Where are the metaphors to guide us into the next season of our becoming?

By reflecting upon our life experience, we can turn even our regrets into wisdom. Our insights may not only nourish the seeds of the future, they might also encase the spouts of an unlived life still longing to come to fruition, over the summer.

Such a sprouting is possible here in the desert, where people hibernate in the summer, hiding from the heat, and fall signals a second flowering for plants that have gone dormant.

Returning home, I pause beneath the non-native flowering tree in my front yard that is "raining," spurting droplets of water. These slants of moisture within the branches show themselves only at just the right angle in the early morning sun's light and rising heat. Is the tree weeping, like Jesus?

Surely we're beyond that paradigm preserved in another Garden, on the eve of atonement for what happened in *Eden*.

Perhaps our planet itself holds a clue to the new story. Here on Earth, where the cosmos became chlorophyll, we humans consciously hold our place in the long line of life moving through honeysuckle, trees, doves, and hummingbirds.

Red Spring Enchantment

I had 'found' this special place at the end of my interim year; back then it was nothing but a faint trail winding through a few trees and around outcroppings of rock. On one of these I'd accidentally come upon petroglyphs…there was no sign to suggest their presence, so seeing them came as a complete surprise….and a deep delight. Within this spring fed oasis, I would spend time amid the symbolic life of earlier peoples.

These several years later, however, the area has become a major attraction; a boardwalk tries to keep visitors off the fragile meadow and a respectful distance from the petroglyphs. But someone had climbed over the rope and scratched ALICE across one panel, totally defacing it. And the cleanup effort to sandblast away the name erased the ancient symbols as well.

Long ago, I let go of the doctrine of Original Sin, but spending time in our public lands has given me a new appreciation of why all organized religions have tried to place limits on human misbehaviors. Dismayed, I leave the scene of children running across the meadow behind the signs that tell them to keep off, while their parents ignore their behavior.

Fleeing, I climb into the silence of red sandstone to find a boulder-created slant of shade from where I can see the larger narrative that begins with the warm inland sea of

600 –250 million years ago, when this area lay near the equator: trilobites, crinoids, brachiopods told that part of Earth's Story.

After the sea retreated came the braided streams of 250 –180 mya, leaving mud flats, with sandy silt deposited to become sandstone and shale: alligator-like reptiles, large amphibians, and crayfish made up that part of the Story.

Then the giant sand dunes of 180–60 mya formed as the climate became more arid and winds created huge dunes 2,000 feet deep that compressed and cemented into the red sandstone that defines the desert southwest where once small dinosaurs roamed. Next, the woodlands of 1.8 mya –10,000 years ago tell of a cooler wetter landscape that supported mammoth, mastodon, horse, and giant ground sloth, and bear….and possibly people.

Finally there's the arid desert from 10,000 years ago to today, with the climate becoming hotter and dryer and replacing the by-then 'extinct' species with the plant, animal, and human species now inhabiting this desert with its isolated mountain springs.

Now, in today, there seems to be a stray Petroglyph on a rock a little distance away; I study its strange geometric shape. Does anyone else know this is up here? Or is it so insignificant that no one has bothered to mark it with a sign, like those below. I shall share this 'secret' only with respectful friends.

On another rock nearby, a chuckwalla is doing pushups, sunning itself in light from the beginning of Time that is just now reaching us both, gathering us together in the magic of Light become Life. As I hear the pulse of the universe beat through my veins, gulp in the fluidity of air, and feel the solidity of earth beneath my bones, I'm so enchanted I'm sure I can fly.

It is precisely this experience that mystics from every faith tradition try to describe….with about as much success as I am having here now. Yet this sense of completeness is the birthright of every human being, and is, perhaps, the reason we have emerged out of stars and onto this planet at all.

So when we extinguish our species through our shameful behavior, will that left-over dinosaur on the other rock reclaim the world? And the Universe's self-reflective awareness vanish?

A Desert Is Where

The desert is where Light is so intense and unrelenting that you can believe it is coming from the beginning of Time, feel its heat fire your cells into life, its energy fuel your imagination.

And so you strike a candle-flame for contemplation, let its light become your enlightenment, its glow invade your whole being with Being and spark your further Becoming.

The desert is where layers of deposited Earth are lifted up into your epoch and you know that the planet's story is playing out in your own flesh and bone, and that its heartbeat is your own.

And so you pick up a pen, extending your awareness of what wants to come through you, let the DNA of your fin, flipper, limb, wing, claw, paw become hand have its say.

The desert is where Water is so precarious it is precious and its pulsing through your body leaves you gulping for more as you shift between dwindling fossil aquifers and current snow melt.

And so you become as water, imbibing the nutrients from the great soup of life by reading the record of those here before, and gleaning the wisdom from all that's here now.

The desert is where Air that's unfiltered by the moisture from oceans and trees makes you acutely aware of the interchange of your own breath with the breathing of all other life forms.

And so you accept that your breath passes over vocal cords and that you've been given a voice to speak up for all, bring every living thing into the planet's conversation.

A desert is anywhere your ego-self becomes an Eco-Spiritual Self.

Graced Moment

If one day we get to stand at the Pine Creek trailhead, looking down the steep gravel slope, and think we see nothing.... utterly nothing...but a desolate desert landscape, let's head down that trail that will take us to remnants of yesterday's Ice Age, slip on the gravel until we come to the smoother multi-colored layers of several of Earth yesterdays.

Whole eons slide under our feet in the grey dirt becoming brown becoming beige becoming yellow becoming red before becoming the soft pink of smooth sand as the trail levels out.

A rare rain may make colors more vivid as we head across the arid landscape that was once a water-fed habitat, detour to sniff the trunk of the Ponderosa Pine...vanilla? Butterscotch!

As we pick our way across a seasonal stream on stones polished throughout yesterdays beyond comprehension, we'll salute a thick stand of willow as it puts forth fur into today.

We'll pause to watch a hawk ride a thermal in circles. As this leftover dinosaur calls out, a small private airplane comes humming into view, and millions of yesterdays become one day.

A hollowed out tree hugs us, sending us forth into a tomorrow co-created by yesterday and today simultaneously, and we are caught in a warp of time that leaves us breathless as we scramble back up the steep slope of our own era, come up out of the desert that is anything but *deserted*. This graced moment in time makes us weep, steeps us in gratitude.

About the Author

For Gail Collins-Ranadive, writing has always been the best way to stay centered and make sense of life's experiences: from being a nurse to earning a private pilot's license; from visiting with in-laws in India to spending time with a friend living in Nicaragua before and after the revolution there; from earning a degree in Peace Studies as a military spouse to lobbying for federal funding of the U.S. Institute of Peace; from creating a women's writing workshop as part of an M.F.A. in Creative Writing to winning a grant to publish it in book form; from earning an M.Div. to doing ordained interim ministries all across the continent; from growing up the oldest of eight to mothering two daughters to becoming a grandmother of four granddaughters and one grandson. An Easterner by birth, she currently spends winters at her home in Las Vegas, summers in her partner's home in Denver....always writing, writing, writing.

Acknowledgements

I am deeply grateful for these companions along this journey: Theo Small and Antioco Carrillo, Kathy Utiger and Pauline Antonakos, Colleen Hall-Patton, Megan Rice, and Betty Scott. Thank you also to members of the Sun City Writer's Group, Nevada Friends of Paleontology, The Tortoise Group, Occupy Las Vegas, and the Southern Nevada Chapter of the Sierra Club and Stillpoint Center for Spiritual Development.

At Homebound Publications we recognize the importance of going home to gather from the stores of old wisdom to help nourish our lives in this modern era. We choose to lend voice to those individuals who endeavor to translate the old truths into new context and keep alive through the written word ways of life that are now endangered. Our titles introduce insights concerning mankind's present internal, social and ecological dilemmas.

It is our intention at Homebound Publications to revive contemplative storytelling. We publish full-length introspective works of: non-fiction, essay collections, epic verse, journals, travel writing, cookbooks, skill/trade books, and finally novels. In our fiction titles our intention is to introduce new perspectives that will directly aid mankind in the trials we face at present.

It is our belief that the stories humanity lives by give both context and perspective to our lives. Some older stories, while well-known to the generations, no longer resonate with the heart of the modern man nor do they address the present situation we face individually and as a global village. Homebound chooses titles that balance a reverence for the old sensibilities; while at the same time presenting new perspectives by which to live.

CPSIA information can be obtained at www.ICGtesting.com
Printed in the USA
BVOW05s1751120514

352912BV00001B/9/P